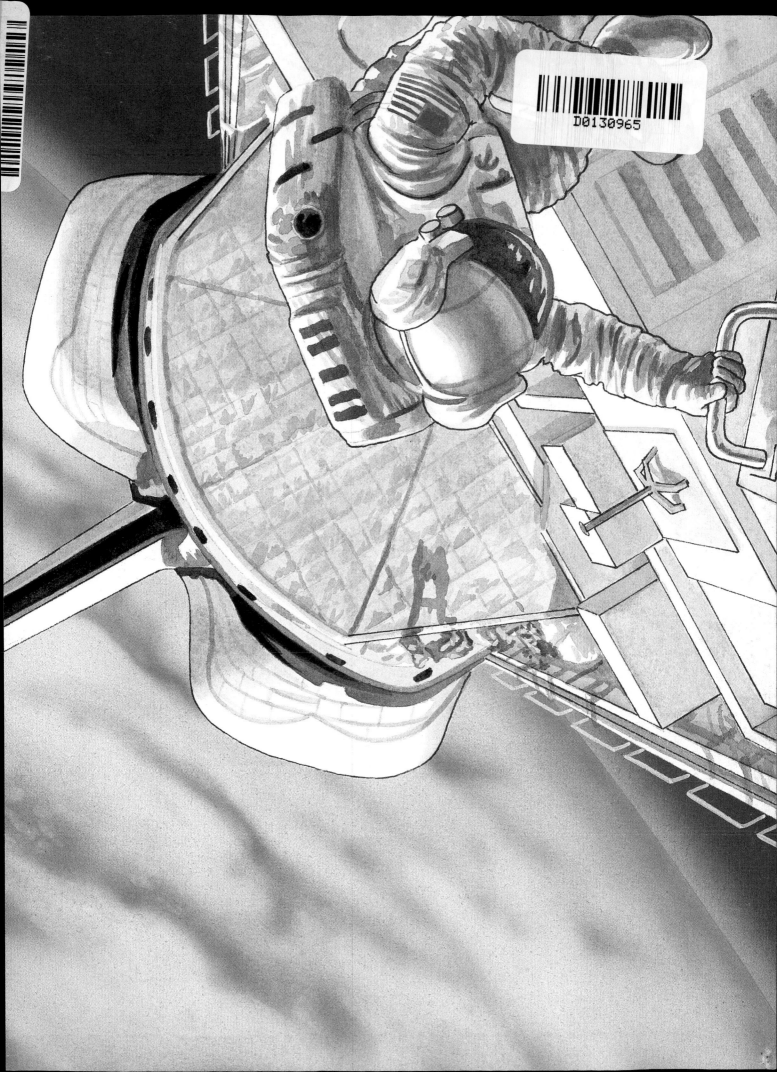

To Elizabeth, Stephen, Rebecca, Rick, and Cindy, and all of the other 21st century stars!

The Golden Book of Space Exploration

By Dinah L. Moché, Ph.D.
Illustrated with
full-color photographs
Paintings by Tom LaPadula

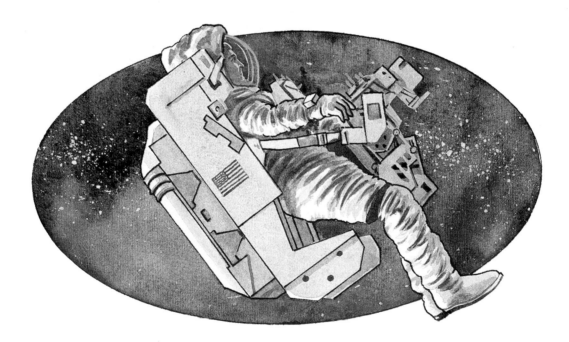

Mark R. Chartrand, Ph.D., Consultant, Vice President, National Space Society

Photographs and artists' conceptions on pages 7, 10, 14, 16, 18, 24-25, 30-31, 32-33, 34-35, 36-37, 38-39 courtesy of National Aeronautics and Space Administration; photograph on page 13 courtesy of Teledyne-Brown Engineering; photographs on pages 22-23 courtesy of Sovfoto.

The author wishes to extend special thanks to: Rebecca Kahlenberg, researcher; Elizabeth K. Rozen, M.D.; Ernest Holzberg, Esq.; Sharon L. Denver: Boeing; Gregory D. Farmer: General Dynamics; Susan Vassallo: Grumman Corporation; Lee D. Saegesser, Althea Washington: NASA Headquarters; Mark S. Hess, Diane Stanley, Peter W. Waller: NASA Ames Research Center; Mary Beth Murrill, Jurrie van der Woude: NASA Jet Propulsion Laboratory; Charles Borland, Mark Craig, Billie A. Deason, Kari L. Fluegel, Michael Gentry, Phyllis Grounds, Vicki Hawthorne, Jerry Homick, Stephen A. Nesbitt, Jennifer Noelka, and Lisa Vazquez-Morrison: NASA Johnson Space Center; Jerry Berg, Margrette Dickerson, Jeffrey S. Ehmen, Jim Sahli: NASA Marshall Space Flight Center; B.J. Bluth, NASA Space Station Program Office; Nicholas Johnson, Teledyne-Brown Engineering; Craig Dunn, Suzanne Hefner, Lee Sentell, and Pat Saucier: U.S. Space Camp.

A GOLDEN BOOK · NEW YORK
Western Publishing Company, Inc., Racine, Wisconsin 53404

Dreams of Flight

Do you ever look up at the sky and wonder what's out there—what exciting worlds are beyond the Moon and stars? Do aliens live there? Can we?

People have always wondered these things. Some worshipped the Sun and Moon. Others dreamed of flying to the stars. But how?

A famous Greek myth tells of a man named Daedalus who copied birds. He and his son Icarus were locked up behind high walls. To escape, Daedalus made wings of feathers and wax. Off the two men flew. Icarus had fun! He flew higher and higher until he went too high. The Sun melted his wings. He fell to the sea and drowned.

A Chinese legend says an official named Wan-Hoo tried to rocket to the Moon. He sat on a seat with forty-seven rockets underneath. Forty-seven servants lit all the rockets at once. When they exploded, he was killed.

Later fantasies had space heroes in flying chariots, electric machines, and hot air balloons. A super-long cannon blasted them off in Jules Verne's great tale *From the Earth to the Moon*.

Exploring the Unknown

What really happened to the first human in space? On April 12, 1961, Soviet pilot Yuri Gagarin lay strapped into a tiny spacecraft at the tip of a giant rocket. Engines fired. Flames and smoke shot out. The rocket and ground trembled. Liftoff! He roared into the sky. Viewers stared up in awe. Would he come back alive?

What a risk! Space has no air, water, or food. It looks black and feels deadly cold, except near blazing-hot stars like our Sun. Dangerous radiation shoots everywhere. Micrometeoroids, or bits of rock, whiz by like bullets.

Astronauts and objects are weightless in space. Scientists call this state microgravity.

Microgravity affects your body and mind. Blood shifts upward from your feet toward your head, so your legs and waist get thinner while your arms and face puff up. You get the sniffles. You grow a bit taller. Microgravity can make you nauseous and dizzy. It weakens your heart, bones, and muscles.

Despite all the dangers, Yuri Gagarin survived! He radioed greetings to the countries below as he flew once around Earth. After one hour and forty-eight minutes, he landed.

Daring men and women have been exploring the unknown in space ever since. They believe space travel will make life better on Earth.

In space, people and things float unless they are secured. It can be fun after you adjust to it. You can do amazing acts— somersault, juggle, lift a friend with just a finger.

Rockets—the Key to Space Travel

Getting off Earth is tough. Earth always pulls everything down toward its center. This pull is called gravity. Jump up and you land fast. Throw a ball into the air and it falls. To beat gravity, you need very high speed. Rockets go faster than any other machine.

As you go up from the ground the air thins out quickly. There's no air in space. A rocket can fly where no air exists.

A rocket works like a balloon you blow up. Keep the balloon closed and air pushes outward in all directions. Let go and air escapes out back. Then the balloon shoots forward.

In a rocket, you get this forward push by burning fuel to make hot, high-pressure gases. The back nozzles let the gases escape. Then the rocket is thrust forward. A rocket carries its own fuel and the oxygen to burn it.

The Chinese made the first rockets over 800 years ago. They burned gunpowder in tubes. Many countries have used this type of rocket for glittering fireworks and for war ever since. But the rocket can't be stopped once the gunpowder burns. Traveling on such a rocket would be like riding a bike with no brakes.

In 1898 Russian space pioneer Konstantin E. Tsiolkovsky first suggested rockets for space travel. But, unlike gunpowder rockets, the new kind of rocket he proposed would burn liquid fuel. You could speed it up, slow it down, and stop it.

The Space Age Begins

The first liquid-fuel rocket was built and fired in Massachusetts in 1926 by Robert H. Goddard. It climbed 41 feet, about as high as a four-story building. Goddard built and tested better ones for fifteen more years.

The Germans also tested rockets for many years. Their biggest rocket was the V-2. It had giant fuel tanks. The V-2 carried a ton of explosives for attacks in World War II. It was the grandparent of our space rockets.

Beep! Beep! The world's first spacecraft called home on October 4, 1957, with amazing news. The U.S.S.R. had rocketed a metal ball into space. It weighed 184 pounds and circled Earth like the Moon. Sputnik 1 had a transmitter and batteries. It signaled for three weeks. That's how the Space Age began!

Sputnik 1 zipped around Earth every ninety minutes. An object that circles Earth is called a satellite. It must go at least 17,500 miles an hour to stay up. That forward speed balances the downward pull of gravity. If the speed is slower, gravity yanks the satellite back down. Rockets hurl satellites to the speed they need to stay in orbit.

Sputnik 1 started a great space race. The U.S.A. and U.S.S.R. competed for spaceflight records.

An American monkey named Miss Baker returned alive from a test flight in space on May 28, 1959.

Life in a Rocket

A dog named Laika rocketed into the records on November 3, 1957. She was the first animal in space. She circled Earth in Sputnik 2. How did she feel? "Fine," reported instruments that checked her body. After a week, she was put to sleep automatically, because at that time no one knew how to bring her back alive.

To fly into space and return alive, you need tanks full of air, and heating and cooling units. You depend on fresh oxygen to breathe and on air pressure around your body to keep your blood from boiling. Spacecraft walls must be airtight and meteoroid-proof. There can be no mistakes! If oxygen escapes through a hole, the crew dies.

You also need food and water, a body waste collector, a two-way radio to the ground, and a heat shield or cover for the spacecraft to keep you from burning when you ram into the air upon returning to Earth.

Scientists tackled the problems. They sent more dogs, mice, and guinea pigs on test flights.

When Miss Baker got home, she ate a banana and slept. She lived happily at the Alabama Space and Rocket Center for twenty-five years. You can visit her memorial there.

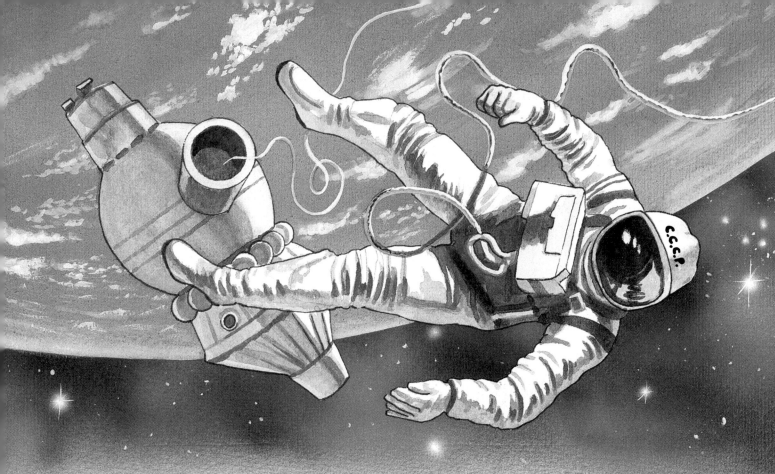

The First Astronauts

Alan Shepard was strapped in a special couch surrounded by equipment in a small Mercury spacecraft on May 5, 1961. He wore a spacesuit that could be filled with oxygen in case the air supply failed. He could barely move. Instruments checked his heart and temperature. He was the first American in space! After a fifteen-minute flight, he splashed down safely into the ocean.

The first American who flew around Earth in space had no time to get hungry. John Glenn circled three times in five hours on February 20, 1962. He ate up there as a test. Can you swallow in microgravity? Easy. Can you keep floating crumbs from fouling up equipment? Hard. Does powdered food taste good? No.

Vostok 6 flew the first woman in space. Soviet cosmonaut Valentina V. Tereshkova circled Earth forty-eight times and parachuted to land, June 16–19, 1963.

Bigger spacecraft carried two or three astronauts. They sat in a section with air. Rocket engines and storage were in a separate section. Soviet Aleksei A. Leonov put on a new spacesuit and went out of Voskhod 2 for twenty minutes on March 18, 1965. It was the first spacewalk! A long hose supplied oxygen, water, and electricity.

To steer a spacecraft, you fire small rockets. Americans Neil Armstrong and David Scott were the first to dock in space. They linked their Gemini 8 to an orbiting Agena rocket on March 16, 1966.

During a period of twenty years each piloted spacecraft flew just once. Now you can see the spacecraft in space museums.

Space Junk

Today satellites are big and heavy. How do we put them up?

Big launchers are made of a stack of rockets, called stages, with a satellite in the tip. The first stage is strongest. It lifts everything into the air. Less powerful upper stages boost the satellite to higher and higher speeds. Each stage drops off when its fuel is used up.

By now there's a lot of junk zooming around Earth—thousands of pieces of old rockets, fuel tanks, dead satellites, as well as billions of bolts, nuts, and paint chips. Chunks collide and make even more junk. It's dangerous and it's expensive to throw rockets away after just one launch. Junk can kill if it blasts into a passenger craft. How can we stop space littering?

The Shuttle Solution

Make rockets that can be used many times! American space shuttles were the first reusable rocket ships. They fly people, equipment, and satellites up and back like a big space truck. A shuttle is 122 feet long and 78 feet across its wings. It has a small crew cabin up front with a big cargo bay behind.

Three main rocket engines are in back. Their fuel is in a separate jumbo tank. It holds enough fuel for 26,000 cars! The main engines and two solid rocket boosters power the shuttle into space.

Shuttles have blasted off and returned safely many times. *Columbia* flew to space first, on April 12, 1981. Next were *Challenger, Discovery*, and *Atlantis.*

Tragedy struck in January 1986. *Challenger* blew up seventy-three seconds after liftoff and the crew was killed. Shuttles were grounded. They were made safer to fly. In 1988, the fleet rocketed back to space. Soon *Endeavour* will join it.

The world's mightiest rocket is the Soviet Energiya. It is a two-stage launcher with boosters added for extra power. It stands 197 feet tall, like a twenty-story skyscraper. Energiya means "energy," and it has plenty. It can put 110 tons, the weight of sixteen elephants, into orbit!

The Soviet shuttle, *Buran,* is strapped onto a monstrous Energiya for liftoff. *Buran* doesn't have any main engines, so it can carry more cargo. It can fly without a crew, by remote control from Earth.

Alert! Europe's shuttle <u>Hermes</u> and Japan's shuttle <u>Hope</u> will fly to space soon. <u>Hermes</u> and <u>Hope</u> are about half as big as the American and Soviet models.

A flight director directs the flight from a mission control center. Flight controllers get reports from the crew. They radio up information and instructions. Each controller has a duty. Capcom (capsule communicator) talks to the crew. The flight surgeon takes care of health. The payload officer oversees science equipment. Others manage spacecraft systems and computers.

Preparing for Flight

Astronauts rely on thousands of people who never leave Earth. At a spaceport, planners figure out where, when, and how to go. Engineers design and test equipment.

At last the crew flies in. Reporters and photographers set up. Astronauts rocket up from the Kennedy Space Center in Florida and the Baikonur Cosmodrome in the U.S.S.R. People and computers at launch control center direct the countdown and blast-off. Excited space fans watch.

Near launch time, teams get everything ready. They service, load, and check out the spacecraft. Workers set it up on a launch-pad. Security guards protect it. Janitors clean up.

Things act strangely in microgravity. Let go of a tool, it drifts off. Turn a screw one way and you spin the other way. Spill water, it forms floating drops. If you go into space, you need special equipment and a lot of training.

You can feel weightless on Earth by floating in water. Astronauts practice work on equipment underwater in a huge pool. An airplane that climbs and dives fast lets you feel weightless for thirty seconds. Astronauts try drinking, eating, washing, and putting on spacesuits in a training jet.

Crews study and practice jobs in simulators for a year before a flight. Simulators are machines that look, feel, and act just like a real spacecraft. Computers tilt and turn them. Astronauts flick a switch, rotate a dial or press a key. They see lights go on, pointers move, and monitors read as if in flight. Finally crew and flight controllers rehearse the whole mission together.

What about emergencies? Astronauts have safety drills for fire, escape on the launchpad, and survival in water and wilderness. And each learns to handle someone else's main jobs.

Astronauts stay fit by exercising regularly. Pilot astronauts fly high-speed aircraft to keep up their skills.

Liftoff!

A space shuttle looks like an airplane, but it's smarter. Its complex computers do flight tasks fast. They answer a thousand questions about flight and spacecraft conditions.

Imagine you're an astronaut. You ride to the launchpad three hours before liftoff. You go up the tower in an elevator, cross a bridge, and enter the shuttle on the mid-deck, where the living quarters are.

You go up a ladder to the flight deck. It has 2,020 controls and displays. You have to know them all! The commander and the pilot fly the shuttle. Three mission specialists are in charge of science experiments and cargo. All strap into seats, facing up to the sky.

The crew, launch control center team, and computers work together. Each part of the shuttle, fuel tank, and boosters must be checked out exactly. Your life depends on them!

The main engines fire. As they roar, you feel the shuttle jerk. The booster rockets burn. You're wearing a flight suit that fills with air to protect you in an emergency.

After two minutes the boosters burn out. The empty cases parachute down to the ocean for pickup and reuse. As you climb faster you feel as if there's a rock on your chest. The force you feel is called 3 G's, as if gravity is pulling three times harder than usual. (The normal pull of gravity is 1 G.)

Liftoff! The shuttle clanks and shakes. Hearts pound. You climb!

Life in a Shuttle

After eight minutes the main engines cut off. The empty fuel tank drops into the air and breaks up. You're leaving Earth's atmosphere. Two smaller engines fire to start you circling Earth. You're in outer space! You're weightless!

Inside the shuttle, the air is normal. You wear a T-shirt, long pants or shorts, and slipper socks. A jacket, clean outfits, and sleep shorts are in a locker on the mid-deck. Pockets hold pens, scissors, sunglasses, and a Swiss army knife when you're not using them. Otherwise, they'd float away.

You open the cargo bay doors to let heat out, and you happily go to work. The shuttle keeps circling Earth every ninety minutes for a week.

When you look out the windows, you get a great view of the world. You see beautiful blue oceans, green forests, brown mountains, and cities whiz by below. You can't see your home, friends, or pets because you're up too high. Every forty-five minutes you see the Sun set. When you're on the dark side of Earth, you see stars.

Camping in Space

The mid-deck is your space camper for weightless living.

A space kitchen unit has hot and cold water and a small oven. More than 125 foods and drinks are packed to stay good without a refrigerator. Tuna in cans and beefsteak in foil packs are heated to kill germs. Macaroni and cheese, and cocoa, are dried, and you add water. Bread, apples, and cookies are fresh. You eat three regular meals a day.

The unit has a space sink on the side. It's covered, with two hand holes. A fan draws dirty water away. You take along soap, washcloths, towels, a toothbrush, and toothpaste.

Every day you work out. A space exerciser hooks to a wall. You can run or jog in place because the exerciser has a belt and shoulder straps.

When your watch shows bedtime, you zip into your space bed for eight hours. The bed is a sleeping bag with a padded board and pillow. A bunk or wall hook holds it still. Eye covers and earplugs give you dark and quiet.

The rest of the time your schedule is jammed with tasks. You deliver new objects to space and pick up old ones for repair. Your space muscle is a huge mechanical arm in the cargo bay. You control it from the flight deck.

Space toilets are in closets. How do you stay on? They have seat belts, foot straps, leg bars, and handholds. When you flush them, moving air instead of water carries body wastes away.

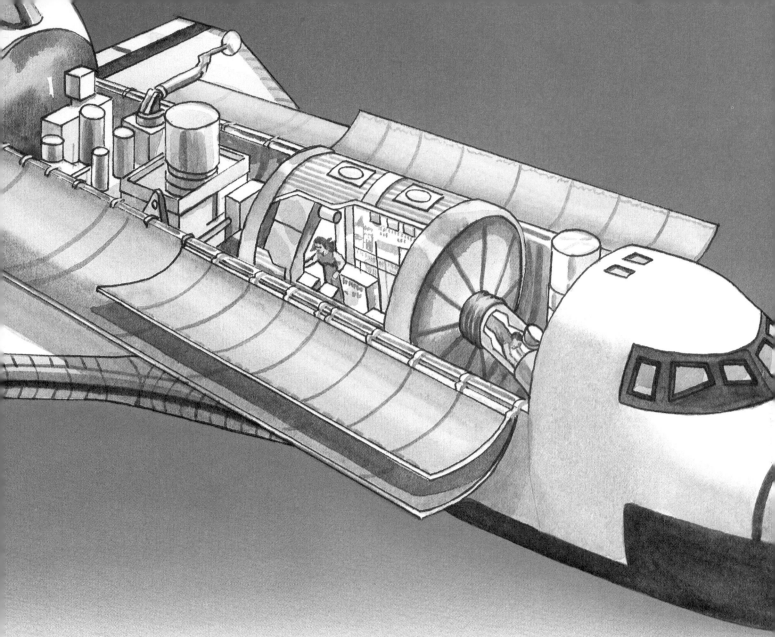

Space Experiments

On some flights crews work in an airtight space laboratory called Spacelab. It has outside platforms to expose objects to space. Spacelab parts fit into the cargo bay in different ways, like Lego pieces.

You enter Spacelab through a lighted, airtight tunnel at the mid-deck. The lab has computers, cages for rats and monkeys, cameras, tape recorders, microscopes, and telescopes. Racks of equipment line the walls.

Space experiments are fun! How do heart, muscles, bones, and blood cells change? You check animals and crew. Do toys work? You play marbles and jacks. What good products can you make? You try new medicines and metals.

Click! Click! You snap fantastic pictures of Earth. People use them to plan use of land and oceans. The pictures show pollution and bad crops, and help locate water, oil, and minerals.

No clouds or air block your telescopes up there. You observe the Sun and stars. You see more than you ever can on Earth!

The Latest in Space Fashions

You wear a spacesuit when you work outside of the crew cabin. The suit protects you from the awful space environment.

The spacesuit has a liner like a football's that holds air inside. Sturdy middle layers keep out deadly cold, heat, and radiation. A tough anti-rip outer layer stops micrometeorites.

You have controls and a display on a chest unit. Your backpack holds enough fresh oxygen, cooling water, and electric power for seven hours. Emergency oxygen is packed below.

The spacesuit weighs 107 pounds on Earth. Like you, it's weightless in space. You put it on alone.

First you put on a urine collector. Next you zip into an undergarment full of cooling water tubes and air ducts. Then you lower both feet into trousers. Bend down and rise up into a hard top hanging on a wall. Connect water and oxygen hoses. Lock top to trousers. Put on a "Snoopy Cap" with headphones and microphones. Adjust oxygen flow. Lock on gloves and a helmet with visor and headlights. Fasten your tools.

In just five minutes you're dressed to go!

When you work in the cargo bay, a tether holds you. The mechanical arm moves you to your equipment.

When you fly off to service equipment in space, a personal rocket set latches to your backpack. You have no ties now. What a thrill!

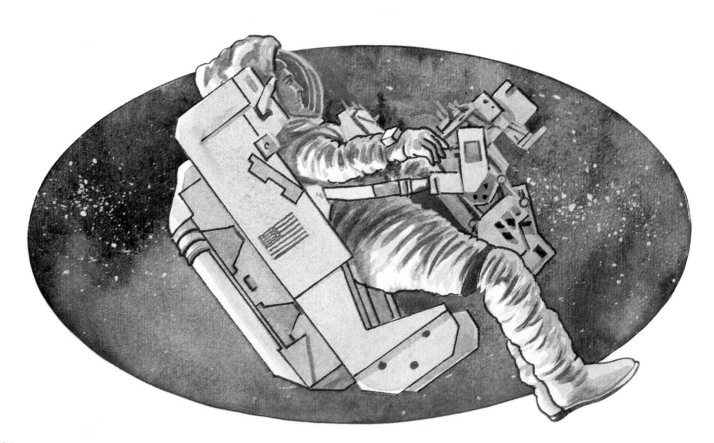

Reentering Earth's Atmosphere

Oh, no! It's time to go home. You follow your checklist.

Close the cargo bay doors. Turn the shuttle backward, tail first. Fire small engines to slow it down. Turn it forward again. Wear your flight suit and put air in it. Ram into Earth's air.

You see the air glow a fiery pink. The shuttle has thick tiles on the outside. They protect you from the burning reentry heat.

The shuttle slows down fast. You feel almost 2 G's on your body. Now the G forces pull blood from your brain to your legs. Your anti-gravity pants squeeze your blood back up, so you don't black out.

You glide to a speedy landing on an extra-long runway. Trucks bring a ground crew over. They cool the shuttle and supply power.

After floating, you feel wobbly back in 1 G. You grip the railing as you exit. You get a medical checkup, rest, and report to scientists. The shuttle is inspected and readied for a new flight.

Alert! Mir has two big rocket engines and thirty-two small thrusters for control. It will stay up for years. Watch for more news!

Space Stations

Someday Mars explorers will be away from Earth for years. What will space life do to them? Will they get homesick or grouchy? How can we prevent harm?

A space station is a place to find out. It's a big satellite that keeps circling Earth. Crews rocket up for jobs that last months. The station is their laboratory, factory, and hotel in space.

In 1971 three Soviets stayed for twenty-three days in Salyut 1, the first station. Then they left in a Soyuz spacecraft. On the way down, air escaped. They died because none wore an air-filled flight suit. Now crews do. Salyut fell into the air and burned up six months later.

Three crews used the first American station, Skylab. Some got space sick. At times they got tense. Checkups showed body changes. All adjusted and did their jobs. A spider name Arabella flew for a test. Could she spin a web in microgravity? She did!

In 1973–74 Skylab crews circled the world a total of 2,476 times. They took 46,000 pictures of Earth and 175,000 images of our Sun.

Soviets set records for long spaceflights in the Mir station, launched in 1986. Mir has an airtight science laboratory and living quarters for six people. Crews enjoy a space kitchen unit, bathroom with shower, sleeping bunks, and fun area.

When people stay a long time on Mir, their muscles get weaker and they lose a few pounds. They get restless up there after a while.

To cheer up, they exercise on a space bicycle or eat fancy foods, fresh fruit, and chocolate bars. On weekends they chat with their families.

Computers help fly Mir and send information to Earth. Special panels turn sunlight to electricity. Six docking ports hold other craft.

Crews taxi up and back in Soyuz spacecraft. One always docks at Mir. They have fire drills and practice escapes in a Soyuz.

Does the crew need more food, water, or fuel? Are they lonely? Robot spacecraft named Progress deliver supplies and mail.

Do they want to observe stars? Craft named Kvant plug into Mir. One has an airtight laboratory and five telescopes. Instruments are exposed to space in its cargo bay.

Mir crews have games, movies, musical tapes and instruments.

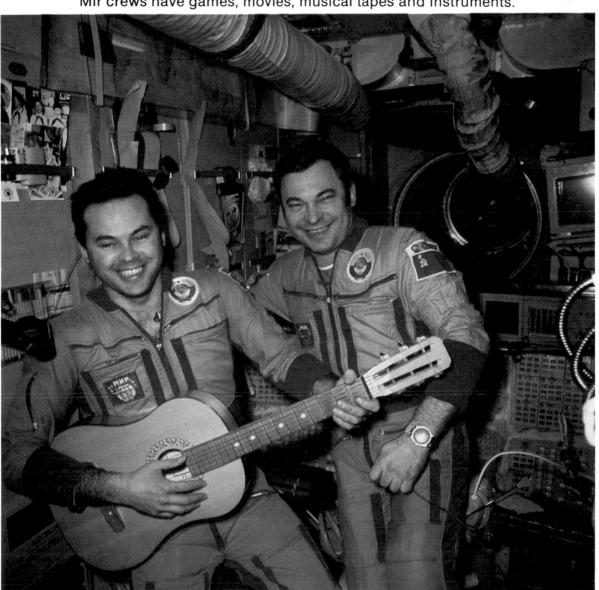

Space stations have medicines and doctor's tools on board for first aid. You can look at a whole library of medical facts on a computer screen.

Building for the Future

Space station Freedom will be like a busy spaceport in space. The U.S. is building it, with Canada, Europe, and Japan as partners.

Space shuttles will fly up crews and parts. The parts will fit together like giant Tinkertoys. Astronauts will build Freedom in space as they circle the world.

First they will make a frame. Then they will attach living, laboratory, and storage units. Solar panels will make electricity. Workstations, outside science platforms, and windowed observation rooms will be added. There will be more additions as time goes on.

Imagine you're an astronaut on Freedom.

You live in a tube the size of a trailer. Walls, floor, and ceiling are all used in microgravity. So the room seems bigger. A crew of eight is on board for three months. How do you stay healthy and friendly the whole time?

To relax, you watch TV and videos or listen to music, you phone home, or you use a computer.

The kitchen has a refrigerator, freezer, and microwave oven. So you eat your favorite foods. Desserts taste good! There's a dining table. Restraints keep you from floating away from it. Cleaning up is easy— you use a dishwasher and trash compactor.

The bathroom has a space shower. In the shower stall, flowing air moves warm water over you. There's a washer and dryer for clothes. Precious water is used over and over again. Waste water is purified on board.

An emergency craft stands by. If an astronaut is very sick or some space junk blasts a hole in the station, the craft takes the crew back to Earth fast.

Like other stations, Freedom will be a space laboratory, observatory, and repair shop. It will also be a launch base for Moon and Mars explorers. They'll build spaceships and pick up fuel and supplies. Then they'll blast off!

Powerful computers will run all systems on Freedom. Robot machines will do many tasks. The crew will be able to operate a long mechanical arm or let it work automatically. It will lift and carry. Its hands will grip and hold.

Free-flying robots will be based at the station. They will make pickups and deliveries and do repairs in space.

Alert! Freedom crews will come and go continuously for many years. Watch the space station grow. Or plan to fly on Freedom yourself!

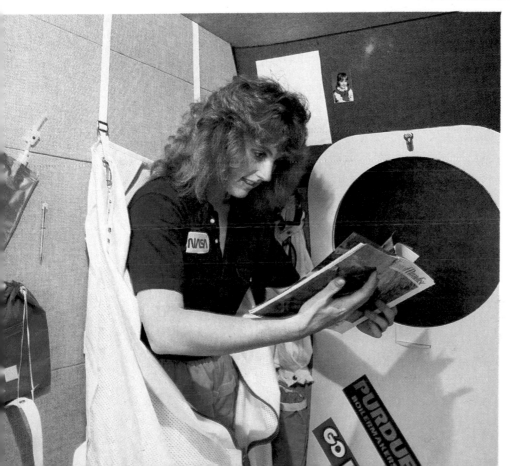

You name it, Freedom has it! You get your own compartment. A sleeping bag hangs on one wall. You put away your clothes and set out pictures on another.

Space Robots

Some space jobs are unusually dangerous. Others take a very long time. Space robots do them!

The robots work alone or by remote control from Earth. They radio information back down. Small rockets are used to guide them.

Each space robot, like each person, looks different. But all the robots have the same basic makeup. Their bodies are lightweight. They use cameras to "see" and computers to "think." They have antennas to "talk" and "listen" to mission control.

Many of the robots circle Earth in space. They are satellites.

Some explore our own world. We still have many questions about it.

Where are natural resources? How can we use them best? Earth-observer satellites whiz over the North and South Poles as our planet spins. They scan the whole globe every two weeks. They beam down images of land features, plants and animals, oceans and air.

What will the weather be? Will a big storm hit us? Every day weather satellites measure air temperature, water, winds, and ice. They snap cloud pictures. You can see some of the pictures on TV.

Can air pollution kill us? Atmosphere watchers keep track of changes up there.

The Sun is a huge, active ball of fiery gases. How do explosions there affect us? Sun observers watch flares that blast electric bits toward Earth.

Communications satellites circle Earth every twenty-four hours. They are 22,300 miles up. They pass TV programs, phone calls, and computer data quickly to people around the world. Tracking and data relay satellites link satellites and mission control. They could relay all the facts in a twenty-volume encyclopedia in one second!

Robot satellites cannot stay up forever. Only our natural satellite, the Moon, always stays up. Robots work for years. If a part breaks, an astronaut may fix it. But all robots finally slow down. There's no way to guide or save a robot after its fuel is used up. Then gravity yanks it back to Earth. It burns up when it reenters the air.

The Far Side of the Moon

The U.S.S.R. and U.S.A. began shooting space robots to the Moon in 1958. All failed at first. The next year was luckier.

Luna 1 got close! The U.S.S.R. launched the metal ball on January 2, 1959. With its instruments and antennas, Luna 1 weighed 361 pounds. Space robots that go to the Moon or beyond are called probes. Luna 1 was the first probe to escape Earth's gravity.

The first probe to actually hit the Moon was Luna 2. It crashed there on September 15, 1959.

When we look at the Moon in the sky, we see the near side. The far side is always turned away from Earth. Luna 3 took a camera there on October 4, 1959. It sent back the first pictures of the far side.

In 1961 President John F. Kennedy told Americans, "I believe this nation should commit itself to achieving the goal, before this decade is out, of landing a man on the Moon and returning him safely to Earth."

Wow! What a challenge! We can't see what the Moon is like from Earth. It's 240,000 miles away. In 1961 no one knew if it had safe landing sites. No one even had a Moon rocket for astronauts.

Americans took the challenge. The project was named Apollo. In time, 400,000 excited workers took part.

You can't hit the Moon by aiming right at it. The Moon always moves. It goes all around Earth every twenty-seven days. You have to aim a probe so that it meets the moving Moon. It's like throwing a basketball to a friend who's running across the court.

Landing is tricky. The U.S. probe Ranger 7 sent us the first close-ups as it hit the Moon on July 31, 1964. The Moon has high mountains and deep craters. Only smooth ground is safe. But the Moon always spins. It turns all around every twenty-seven days. So good landing spots don't stay put.

Scientists must figure exactly where a good spot will be far in advance. Mistakes kill. Computers help a lot in finding the answer. They can add more numbers in a second than a person can in a lifetime.

On February 3, 1966, the U.S.S.R. probe Luna 9 landed safely! It sent back pictures for three days. Later, U.S. Surveyor probes landed, took pictures, and tested the soil.

Then scientists chose a Moon landing site for astronauts.

The Apollo spacecraft was ready. A command module housed the crew, their food and spacesuits, and flight controls. A service module had rockets, fuel, air, and power supplies. A lunar module would land two men. The whole load blasted off to the Moon on a Saturn 5 rocket. It was the biggest rocket ever built.

"One Giant Leap for Mankind"

Apollo 1 was tested on the launchpad in 1967. Disaster struck. Fire flashed inside. The crew was killed.

After that, planners worked to make Apollo flights safer. Crews began test flights in 1968. Apollo 7 circled Earth. Apollo 8 went around the Moon. Apollo 9 flew the Moon lander around Earth. Apollo 10 took the whole Moon trip except for the landing. Each tryout was a success. Apollo 11 dared to go for it!

The historic launch was set for July 16, 1969. A million people went to see it. Over 500 million more people watched it on TV worldwide. Saturn 5/Apollo stood thirty-six stories high. Commander Neil Armstrong, Edwin "Buzz" Aldrin, and Michael Collins sat at the top.

Saturn 5/Apollo weighed six million pounds. As the countdown ended everyone stared up. Liftoff! The crowds clapped and cheered.

Three days later Apollo 11 circled the Moon.

On July 20 Armstrong and Aldrin put on spacesuits and went into the Moon lander *Eagle*. They flew down to the surface. The world waited anxiously.

"The *Eagle* has landed," Armstrong called down.

"We're breathing again. Thanks a lot," replied Earth.

Armstrong opened the hatch. His boot set upon the gray dust. "That's one small step for a man, one giant leap for mankind," he said.

The Moon's surface has no grass, plants, insects, or animals, just dust, pebbles, and rocks.

Six crews followed Apollo 11 to the Moon. Apollo 17 was last, in 1972. All together twelve men spent three hundred hours on the Moon. The last three teams drove around in Moon cars.

Soon Aldrin joined Armstrong. The world watched them live on TV. The Moon has no air or water. Backpacks held air, cooling water, electrical power, and two-way radios for the astronauts. Supplies would last four hours. The pair set up experiments. They collected soil and rocks. Meanwhile Mike Collins kept circling the Moon in the command module.

Taking a Moonwalk, you hop like a bunny. The pull of the Moon's gravity is very weak. You feel strangely light.

Armstrong and Aldrin finished their work. Then they rejoined Collins. Apollo 11 flew home. Its service module was cast off in space. The command module splashed down into the Pacific Ocean on July 24. President Richard M. Nixon met the heroes on a pickup ship.

Later Apollo crews took thousands of great pictures. They got 837 pounds of rocks for study. Instruments they set up sent facts for years.

It's time to build a Moon base! U.S. President George Bush declared in 1989, "For the new century…back to the Moon…to stay. And then… to Mars." His challenge calls for new ideas and brave explorers. Bases beyond Earth will use new rockets, shelters, land vehicles, and power plants. At first the bases will get all supplies and equipment from Earth. In time they will fill their own needs.

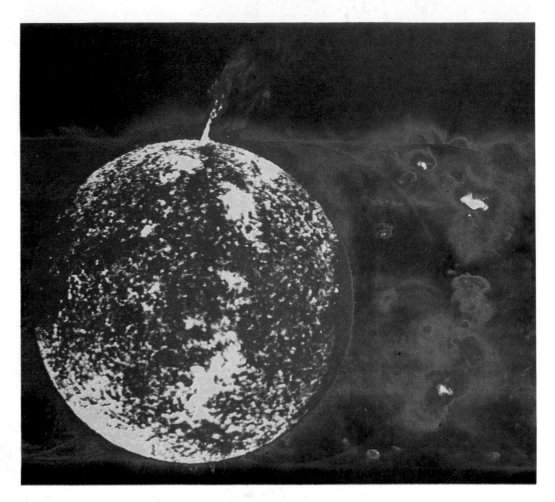

The Solar System has one star, our Sun.

Our Neighboring Planets

Our Earth and Moon belong to a vast space group called the Solar System. There's a lot of room for human settlers to build colonies.

The Solar System has eight planets besides Earth. Most of them have moons. There are also swarms of rocks of different sizes, and huge iceballs.

Everything in the Solar System keeps moving around the Sun. Don't worry—the planets never collide. The Sun's awesome gravity keeps each one in its own orbit, like a train on a track.

The ancients saw five planets in the sky. They named them after their gods and goddesses. Mercury, Venus, and Mars move nearest to Earth. Jupiter and Saturn are huge. From Earth they all look like bright stars.

Space explorers are curious about what the planets are really like. They want to see for themselves. But the planets are very far from Earth.

Venus comes closest—25 million miles away. That's 100 times farther away than our Moon!

Venus was named for the Roman goddess of love and beauty. The planet looks gorgeous in the sky. It sparkles! But when probes began to fly there in 1961, they discovered that it's really horrible.

At 900°F, Venus is the hottest planet. It always has a thick yellow cloud cover. The air is poisonous and crushing. It is 90 times heavier than our air.

Soviet Venera probes landed on Venus and sent back pictures. They show sizzling orange-brown rocks. The heat and pressure killed the probes fast. Human explorers won't go!

Venus is the same size and weight as Earth. How did it become so unlivable? We need more facts. Understanding Venus can help us keep Earth safe.

Venus has rocky plains, dry seabeds, and high mountains. There are no plants or animals.

Mercury is also rocky, dry, and lifeless. It is the closest planet to the Sun. It gets burning hot. The pictures show many craters. Mercury has no air to stop meteoroids, so it gets hit endlessly.

Mariner 10 flew by Mercury and took pictures three times in 1973–74.

The Mysteries of Mars

Mars looks sparkling red in our sky. It excites people more than any other planet.

Probes began to send back facts about Mars in 1965. The whole planet is a rocky desert, so its dry riverbeds are startling. Mars has big craters, deep canyons, and dead volcanoes. Some ice is always at the north pole, but there's no flowing water.

Mars must have been somewhat like Earth, long ago. It had rivers. Perhaps it also had living things.

Two U.S. Viking robots landed to look in 1976. Both Vikings tested scoops of soil for tiny living things. They didn't find any. But they only checked two sites. They couldn't move to other parts of Mars.

The Vikings gave three million weather reports. Mars is usually freezing cold. The air has just a bit of moisture. It never rains on Mars, but winds blow sand wildly at times. Mars air is poisonous. It's so thin our blood would boil there.

Mars still has many mysteries: Were living things ever there? Is a lot of water frozen underground? A few features look like faces and pyramids. What shaped them?

Space fans worldwide want to know. They aim to build a Mars base together. Will you help?

The Viking robots took 4,500 great pictures. None show plants or creatures, just sandy red plains with a pink sky.

Each Viking had an orbiter and a lander. Today the Viking 1 lander belongs to the U.S. National Air and Space Museum. It's the first museum exhibit on another world!

First more space robots will explore Mars. Some will circle the planet and get facts from space. Others will land and roam on wheels. They'll test soil and rocks. They'll bring samples home for study.

Meanwhile, planners prepare for piloted Mars flights.

The first crew must take along all the air, water, food, and tools they need to live. They'll also need a Mars lander, shelters for living and working on the surface, spacesuits for protection outside, a Mars car, and rocket fuel.

One idea is to launch a cargo ship first. At Mars it would circle the planet. A crew ship would be launched later. After eight months it would link up with the cargo ship. Some astronauts would go into a lander and fly down to the surface. The rest would work up in the main crew ship.

How would the base grow?

New crews recycle water and wastes. They grow food in greenhouses. They make breathable air, building supplies, and rocket fuel from Mars materials. And they set up electric power plants.

Mars has two small rocky moons named Phobos and Deimos. Both seem likely to be good sources of water and rocket fuel. The moons have hardly any gravity. If you weigh 60 pounds on Earth, you'd be lighter than a Ping-Pong ball on Phobos. So robots and crews could come and go easily. A Mars base could get supplies from them. In time the base would be independent!

An enormous old storm on Jupiter, the Great Red Spot, could swallow up a few Earths.

Lightning Bolts, Icebergs, and Windstorms

Once in 176 years the four biggest planets move into the same part of the sky. Probes can visit them quickest then. The most recent lineup was in 1977.

U.S. rockets launched two Voyager probes that summer. Voyager 2 travels 800,000 miles a day. It flew by the four giants and their moons. Voyager shot pictures, made measurements, and kept going. It has never stopped.

Colorful clouds swirl around the giant planets. Strong winds blow. There's no place to land. These planets are mostly made of gases.

Jupiter is bigger than all of the other planets together. Lightning bolts flash in the cloud tops. Thin rings made of dust particles whiz around. The flight to Jupiter took two years.

The planet has sixteen known moons. Ganymede is the biggest moon ever seen. Io's live volcanoes make it look like a giant pizza. Europa has a glacier cover.

After four years, Voyager flew by Saturn. Dazzling rings loop the planet. They look solid in our telescopes. Voyager's radio signals flew through the rings! They are not solid but are made of ice particles and icebergs as big as houses that zip around the planet.

Saturn has seventeen known moons. Titan's thick orange atmosphere is extraordinary—it has some of the same kinds of molecules that make up living things on Earth. Perhaps Titan has tiny living things! Mimas has a huge crater that was formed when the small moon was nearly shattered in a crash long ago.

Voyager took nine years to reach Uranus. This planet has a weird tilt! Uranus is the only planet whose axis points toward the Sun. Its rings are also unusual. They are made of bits and chunks of dark rock.

Uranus has fifteen known moons. Ariel is the brightest. Umbriel is the darkest, with a lot of craters. Miranda has deep canyons, high mountains, and a strange feature shaped like a "V."

The last flyby on Voyager's trip was of Neptune. It took twelve years to get there. Before Voyager left Earth, scientists gave it instructions for the whole tour. They sent updates along the way. But Voyager had to help itself in an emergency. An SOS from Neptune would take four hours to get here.

The planet has eight known moons. Icy Triton looks bumpy, like orange peel. It has ridges, craters, and frozen lakes. This bright pinkish moon has two towering jets of gas. Nereid looks smaller and darker.

There's a wild storm called the Great Dark Spot on Neptune. It's as big as Earth. Feathery white clouds on the Spot's rim keep changing. Neptune's faint rings have strange bright arcs.

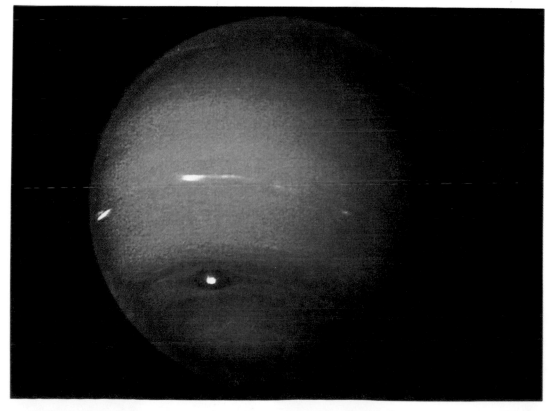

Outbound Robots

The Voyagers took 118,000 pictures. In 1990 Voyager 1 looked back home for the last time. What a great view!

The Sun shines against the blackness of space. When you're way out there, all the planets are in sight. You can spot blue-white Earth between reddish Mars and yellow Venus.

Today both Voyagers are wandering out to the stars. Voyager 2 can call home until the year 2015. Then its signal power will run out. After 21,000 years in the dark, Voyager will near our next closest bright star, Alpha Centauri.

Between Mars and Jupiter thousands of rocks called asteroids always circle the Sun. The largest is Ceres. It is as big as the state of Texas. In our sky it looks like a faint star.

What do asteroids look like close up? What are they made of? Could we get valuable raw materials from them in the future? Outbound robots will take asteroid pictures as they fly by. Future robots may visit asteroids.

No outbound robot sets out with enough energy for its whole trek. It gets speed boosts along its tricky path.

The Voyagers are ready to meet aliens. Each has a golden record in a gold case. Sounds include a kiss and a baby's cry, greetings in fifty-five languages, and the world's best music. There are also 118 pictures of Earth and its people.

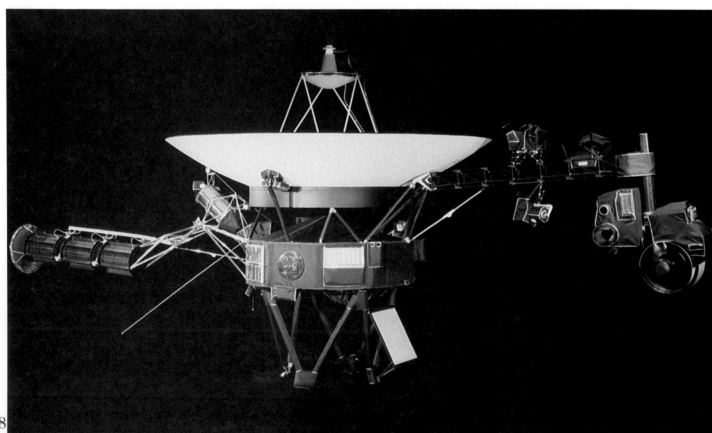

Galileo was sent to Jupiter from space shuttle *Atlantis* in 1989.

The U.S. robot Galileo loops around planets like a ball in a gigantic pinball machine. It heads to Venus, then flies twice by Earth. Each planet's force of gravity speeds Galileo onward.

After two years, Galileo will zip by asteroid Gaspra. Gaspra is as big as New York City. It will be the first asteroid seen close up.

After six years, Galileo will near Jupiter. It will split into two parts called orbiter and probe. The orbiter will be the first robot to circle Jupiter. The probe will be the first into Jupiter's atmosphere.

Jupiter's mighty gravity will pull the probe down. The probe will hit the swirling clouds at 100,000 miles an hour. A huge force will slow it fast. While sinking, the probe will beam facts up to the orbiter. The orbiter will pass them back to Earth.

After an hour, heat and pressure will kill the probe. The orbiter will go on exploring Jupiter and its big moons for two years.

The U.S.-European robot Cassini will near Saturn in the twenty-first century. It gets speed boosts from the force of gravity at Earth and at Jupiter. Near Saturn, Cassini will split into two parts. A probe will dive into the atmosphere of the moon Titan. It will do tests and snap pictures of the surface. After three hours it will land. If it lives, it will send a brief report. An orbiter will keep looking at Saturn and its moons for four years.

Star Light, Star Bright

The stars in the sky are huge, hot suns. They look tiny because they're trillions of miles away.

Light speed is the fastest speed in the universe. Sunlight zips to Earth in eight minutes. But the other stars are so far away that many years go by before starlight reaches Earth. Starlight takes 782 years to get here from the North Star. So when we see stars, we're looking back in time. We see the North Star as it looked 782 years ago.

From the ground we look at stars through Earth's air. We can see 2,000 stars at best. Powerful telescopes help us see more. But the air always blocks a lot. You get the best views of stars from space.

Satellites fly telescopes above our air and clouds. The U.S.-European Hubble Space Telescope sees many more stars than the biggest telescope on Earth can. It sees the most distant stars known and all the way back to the beginning of time!

The Hubble Space Telescope is as big as a bus, with an "eye" that is eight feet across. The satellite also has two cameras and three starlight analyzers, and takes planet pictures that look like close-ups. Its detectors could pick up the glow of a flashlight on the Moon.

Does the Hubble Space Telescope see aliens? No. Any aliens would be too small to be seen at such a big distance. While no creatures can live on a fiery star, aliens could be on planets. Livable planets probably circle other stars like our Sun. The Hubble Space Telescope could spot a big planet.

We search for aliens by listening for a message from them. Our big antennas send and receive coded radio messages between Earth and spacecraft. Smart aliens probably understand the same mathematical codes. They could contact us by beaming a message to Earth.

Today we don't send out messages from Earth to possible aliens. Some people are afraid that aliens who heard from us would invade our planet. It's very unlikely, but not impossible. We can tune in to space without fear. If we get a message, we don't have to tell the aliens.

What's their channel? That's the problem! There are endless possibilities. Luckily computers check millions of channels at once. We haven't heard from aliens yet. But we might today!

Alert! A space telescope works for years. If parts fail, astronauts on a shuttle can help. They fix or replace old parts. Watch for many discoveries!

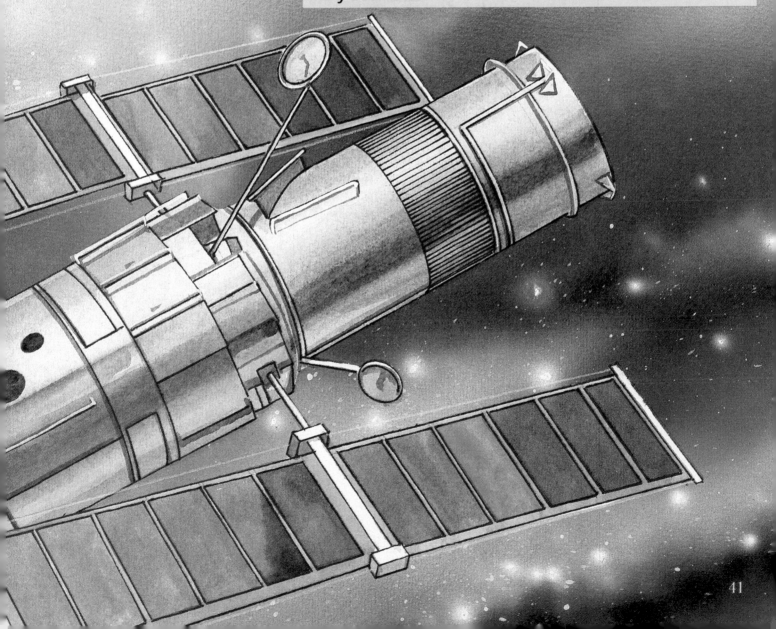

Be an Astronaut...

A space explorer's job is demanding and risky. You need curiosity, skills, and courage. Your rewards are adventures, new friends, and pride.

You can start now by getting to know the stars. Look at them often. Visit a planetarium. Star pictures are projected on the inside of a big dome. You'll see how stars look at different places and times. Look through a good telescope. Often astronomy clubs, schools, and planetariums offer a public stargazing night.

Go to your public library. You'll find space books, monthly magazines, and videos. Some publications list telephone hotlines and computer bulletin boards for space news.

Join a space, science, computer, or rocketry club in your community. You'll meet other people who want to discover new things.

Visit a space center or museum. You'll see rockets and spacecraft plus spacesuits, supplies, and tools that astronauts use in space. Attend a space camp. You'll try a "mission" yourself.

Meet a space author. Go to a talk in your community. Or ask your teachers to invite an author to your school.

... Or Just Talk Like One

Talk like an astronaut. They use acronyms to save time. An acronym is a word made up of the first letters of several longer words.

AFB	Air Force Base
CAPCOM	capsule communicator—person who talks to flight crew
CRT	cathode-ray tube—video screen
ELV	expendable launch vehicle—rocket used just once
EMU	extravehicular mobility unit—spacesuit with life support
ESA	European Space Agency
ETI	extraterrestrial intelligence—smart alien
EVA	extravehicular activity—work outside crew cabin
HST	Hubble Space Telescope
IVA	intravehicular activity—work inside crew cabin
KSC	Kennedy Space Center
MCC	mission control center
MMU	manned maneuvering unit—personal rocket
NASA	National Aeronautics and Space Administration
NASDA	National Space Development Agency of Japan
PLSS	portable life-support system—space-suit backpack
RMS	remote manipulator system—mechanical arm
SRB	solid rocket booster
STS	space transportation system—space shuttle, fuel tank, boosters
TDRS	Tracking and Data Relay Satellite
VTR	videotape recorder
WMC	waste management compartment—bathroom

INDEX

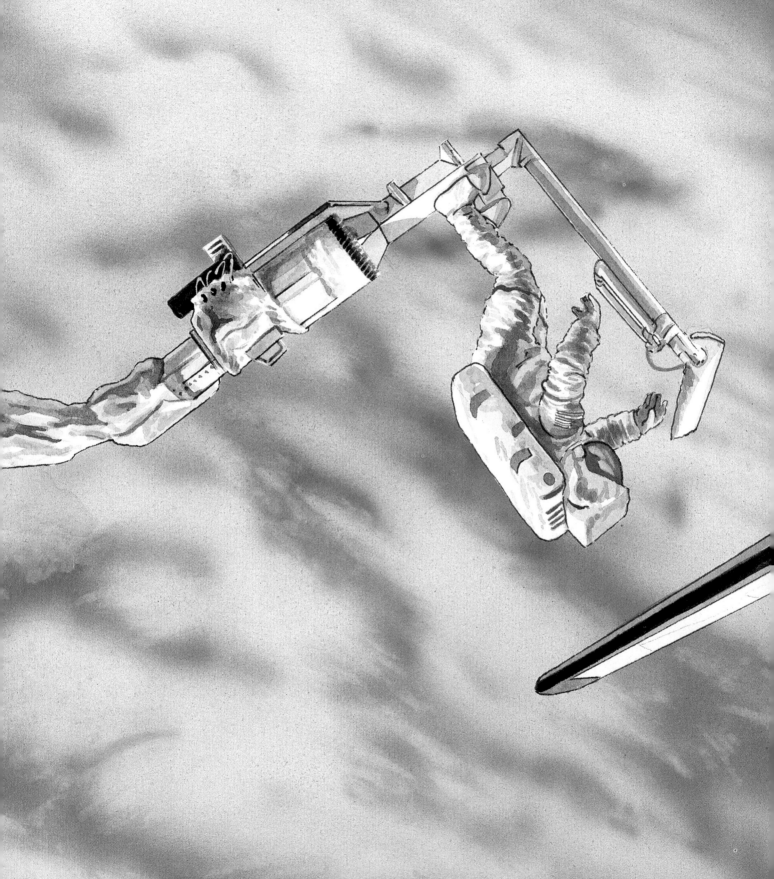